Minitab Statistical
Software

Minitab Statistical Software supports virtually every major
Lean Six Sigma initiative around the world. Minitab's
unparalleled ease of use, comprehensive methods, and
compelling graphics help quality professionals effectively
analyze their data and target the best opportunities for
business improvement.

This book contains references to Minitab Statistical Software.
As a courtesy, readers can download a free 30-day trial at
www.minitab.com.

Contents

Introduction

This guide has been produced specifically for engineers to use as a practical tool in analysing data correctly to solve real world problems. It does not discuss the history or mathematics of the technique except where absolutely necessary to illustrate an important point.

Statistical Process Control is an incredibly powerful tool in the right hands and will help you understand both the variation of data, which always exists, together with the sources of that variation over time.

In this guide a software package called Minitab® Statistical Software is used to analyse data and graph the results. You can download a free 30-day trial at www.minitab.com

Minitab is just one software package that will do this but it is my preferred package simply because I believe it to be the best available – both for ease of use and accuracy of analysis. I have included a number of screenshots taken from Minitab where such screenshots aid explanation.

Traditional teaching of SPC is generally poor in my opinion. This is for two principal reasons:

1. The teaching is carried out by a mathematician or statistician, who blinds the student with complex formulae and makes the subject seem far more complex than necessary. Result: students who will forever decry the technique as too complex or too time consuming to be of any practical use in the real engineering world.
2. The teaching is carried out by a real world engineer, who does not have a sufficient grasp of the subject matter to convey it succinctly and simply. Result: students who don't believe that the technique actually works and therefore steer clear of it in preference of what they know best – looking at data and assuming trends exists (where really none exists) or seeing no patterns or trends when one does exist.

So, work through the text, don't worry it's short, and use Minitab to try out the various examples. You will quickly see just how powerful this technique is in practice and how easy it is to use.

What is Statistical Process Control?

So what is "Statistical Process Control" or "SPC" as it is known in its frequently abbreviated form?

Let's break the term down to enable a better understanding:

S - Statistical, this term is used because we use some statistical concepts to help us understand the data and processes we are analysing.

P - Process, because most engineering is carried out using processes and this is how value is delivered to the customer.

C - Control, this term is used because a process needs to be predictable if we are to make sensible decisions about its control and output. I'll touch on this point in a little more detail later.

When Should It Be Used?

In engineering and science, the most common reasons for using SPC are:

To monitor the stability of a process.

All processes vary, even those that are extremely stable. Despite this many engineers and scientists adjust these processes in the mistaken belief that the process has changed and needs to be "adjusted". Control charts provide a powerful means of assessing if a change in the process is real and requires some management action. One of the main mistakes people make is to confuse tolerances with process variability. **There is no causal link between them**. A process has no knowledge of engineering drawings and specifications!

To determine whether your process is stable BEFORE making improvements.

Adjusting a process in an attempt to change its output is futile. The process needs to be stable or "in control" before such adjustment is made.

To demonstrate improved process performance.

Once a process is stable you may wish to take action to improve it. A control chart can be used to actually demonstrate that the performance has shown a real and sustained improvement.

Main steps when using Statistical Process Control (SPC)

Step 1: Decide what you Intend to Measure

The first step is to select the process output you intend to measure. There are two types:

- Continuous
- Attribute

Continuous means measuring a part or parameter that is continuous in nature such as length, weight, volume or time.

Attribute means dividing parts into good or bad, acceptable or unacceptable. These are clearly counts of whole units.

Wherever possible choose continuous data since it provides much more information about process performance for the same amount of work as collecting attribute data.

Step 2: Qualify the Measurement System

This step is frequently not considered sufficiently when engineers are looking at measuring data. Despite this, it is of the upmost importance because every measurement system has some error associated with it. In many instances the error is large when compared the process tolerance and if this is the case the subsequent analysis and actions will be flawed. In some cases this will also result in MORE process variation rather than less!

This step is generally known as "Measurement System Analysis" or MSA for short.

There are 5 ways to characterise a measurement system. Three of these look at the measured values when compared with actual values – commonly called "Accuracy". The remaining two look at the spread of measurement data – commonly called "Precision".

Let's look at all 5 in more detail:

Accuracy

Stability – if we measure the same sample or part then we need those measurements to be stable over time when measuring the same part. In other words the same part, when measured repeatedly, should measure more or less the same each time.

Bias – commonly called accuracy, this measures the difference between the average measured value and the true real value of a parameter.

Linearity – this is simply a measure of the bias consistency over the range of the measurement device. As an example, consider measuring a part using a 0-25mm micrometer. Ideally we want the bias to as consistent when measuring 25mm as it is measuring 1mm. If it is then it is said to be linear.

Precision

Repeatability – if one person uses the same measurement device to measure a part several times then we would want those measurements to be the same. If they are (within prescribed limits) then they are said to be repeatable.

Reproducibility – If different people use the same measurement device to measure a part several times then again we would want the results to be consistent.

Ultimately we want to make sure the measurement system error is as small as reasonably possible when compared with the tolerances and system variation. You should know how much of the variation in the data arises from the measurement system alone. If it's relatively large then you will not know if the variation in the data is coming from the process or the measurement system itself.

A very reasonable rule of thumb is that the measurement system error should be no greater than one tenth of the overall tolerance or process spread.

To ensure we have a measurement system which is fully capable of measuring a particular part characteristic we must do two things:

1. Carry out a "Type 1 Gage Study".
2. Carry out a "Gage R&R Study"

Let's look at the Type 1 Gage Study first.

Type 1 Gage Study

This is used to identify any deficiencies in the measurement system and simply means that we get one inspector to measure one part repeatedly. This provides an estimate of the gage accuracy and repeatability of measurement. Only when this demonstrates that the system is capable of measuring the part reliably should we move onto the Gage R&R Study.

Here's how to do it in Minitab:

1. Get one inspector to measure one part repeatedly. Let's assume we use a micrometer and we take 30 measurements of the diameter of a part with a nominal dimension of 1.500" +/- 0.2".
   My advice is to take at least 30 measurements to ensure a true estimate of the bias and repeatability. Minitab will expect at least 20 values to be entered.
2. Type the data into Minitab in the first column, C1.
3. Now go to Stat/Quality Tools/Gage Study/Type 1 Gage Study
4. You'll see this entry screen:

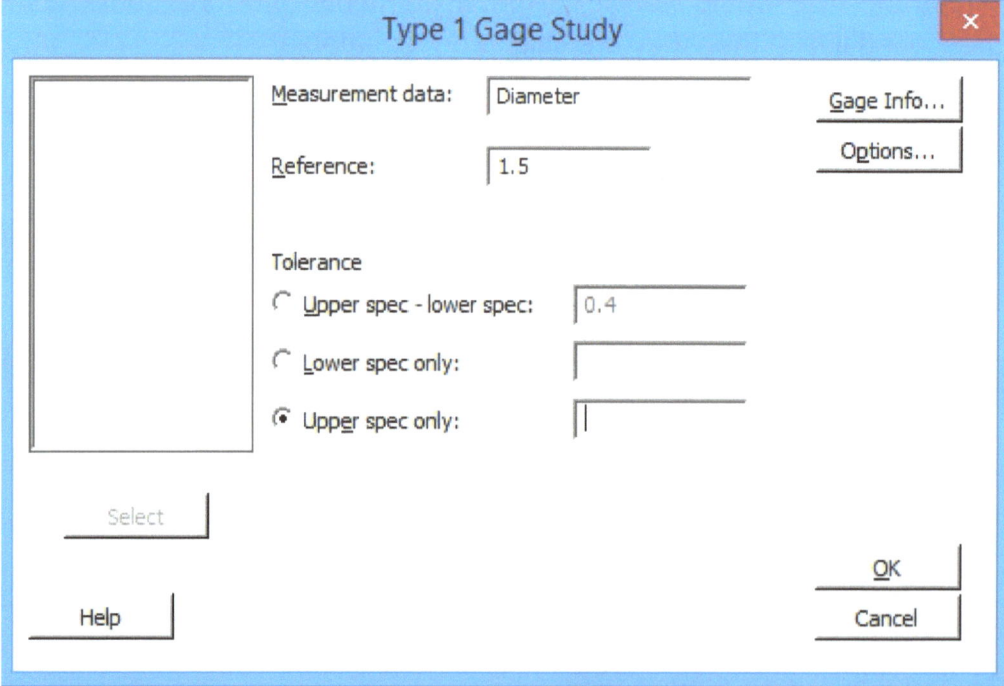

Enter the identity of the measurement data, "Diameter", the reference, which is 1.5", and the tolerance measured as upper spec - lower spec which is 0.4". Ensure that the radio button for this is marked. You can see that the radio button marked in the above image is Upper spec only with nothing entered as a value. Minitab will detect this and return an error message in this case.

You'll see that on the top right of this screen are two buttons marked "Gage Info" and "Options". Let's understand these.

If you click on the one marked "Gage Info" you'll see that it's simply an entry screen to record the gauge name, the date of the study, who is responsible for the study and any other miscellaneous data you wish. Get into the habit of completing this......when you do several studies close together it is easy to mix them up!

If you click on the button marked "Options" you'll be presented with a screen that looks like this:

The "Percent of the tolerance for calculating Cg:" will be pre-filled with the value of 20.0 and the "Study variation" will be pre-filled with the value of 6.0.

Cg provides a comparison of tolerance against measurement variation. Minitab will also calculate Cgk which is a comparison of both tolerance and bias against measurement variation.

Leave both of these at their pre-filled values.

The resolution of the gage can be entered but is seldom required if you have entered a figure for the tolerance. If you do enter a value then Minitab will perform a check to see if the resolution is less than 5% of the tolerance. Ideally it should be!

The "No bias test" option allows you to delete the requirement for a bias test. We want this to be carried out so we leave it unchecked.

Finally you have an option to rename the title on the graph that will be produced. Minitab will place a title automatically based on the header of the column of data we are analyzing. We'll leave it blank and let Minitab choose the title.

Ok, cancel the options screen and we'll go back to the entry screen entitled "Type 1 Gage Study" and click on "Ok"

Minitab will then produce a graph which will look something like this:

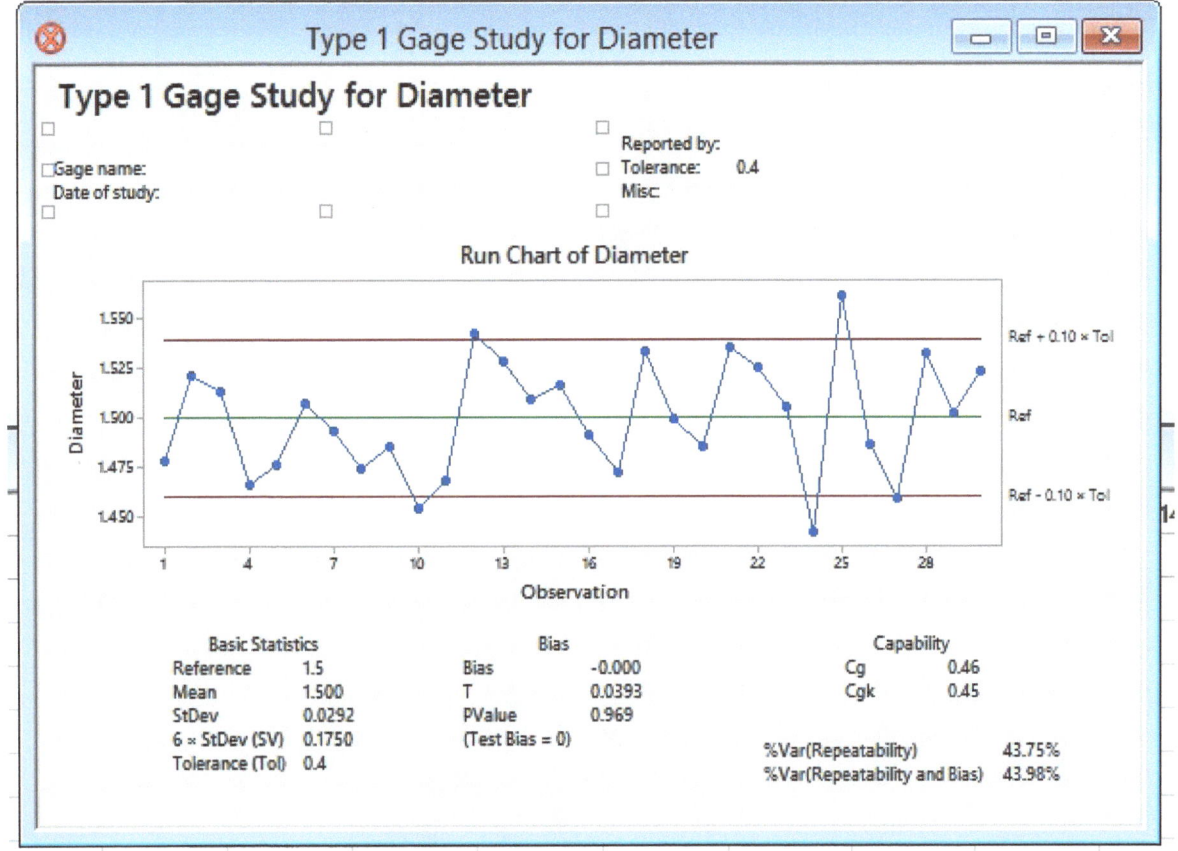

We can now analyze the results. We can conclude the following:

The bias is essentially zero, and we can see this from also looking at the graph and noting that the measured values seem to occur approximately equally both above and below the Ref line of 1.5". If we saw many more measurements either above or below the line then it would be likely that bias was present.

Remember that Cg provides a comparison of tolerance against measurement variation and Cgk is a comparison of both tolerance and bias against measurement variation. The larger the values of Cg and Cgk, the better, since this indicates that the variation due to the measurement system is small when compared with the tolerance range.

It is typical to use a value of 1.33 for the threshold acceptance value and we can see that in this case the Cg is 0.46 and the Cgk is 0.45. We can conclude from this that the variation due to the measurement system is large.

Small values of %Var indicate small measurement variation when compared with range of tolerance and a typical value of 15% is used as the threshold value. In this case these values are both around 44% so the conclusion is that the variation due to the measurement system is too large to be reliable.

I would stop here and take steps to improve the measurement system.

In this case let's assume that we find that the micrometer used was faulty and we sent it away for repair (did the calibration system give us any clues that it was faulty?).

We select a new micrometer and carry out the same experiment again. On analyzing in Minitab we are presented with the following graph:

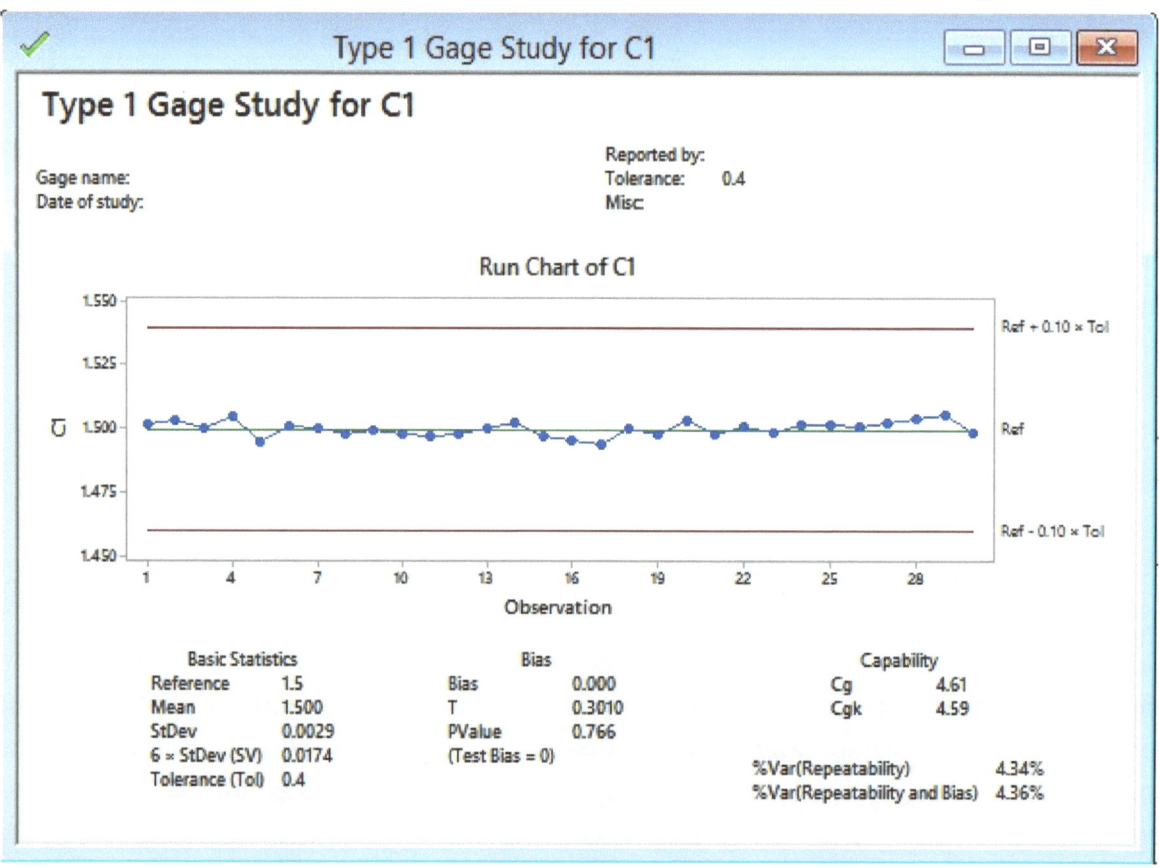

We can immediately see that these results are much better. Both Cg and Cgk are less than the threshold value of 15% and the %Var is around 4%. We can safely conclude that this micrometer can measure parts consistently and accurately.

Note carefully that if we had not carried this assessment it is likely that we would have continued with a Gage R&R Study, the results of which would have been flawed due to the faulty micrometer.

Since the new results are acceptable we can now proceed with the Gage R&R Study.

Gage R&R Study

Again this is used to determine how much of the process variation is due to the measurement system. The difference from the type 1 gage study is that it is likely that there will some variation due to inspectors and between the parts. A gage R&R study therefore includes these sources of error in the analysis.

Here's how to do it in Minitab:

1. Decide:
    a) The number of sample parts,
    b) The number of inspectors,
    c) The number of repeat readings by each inspector

Note Carefully:

- The more parts and repeat readings you take, the higher confidence you will have in the analyzed results.
- Don't be tempted to use "engineers" or other staff to take the measurement. Use the people who usually and normally carry out the measurements.
- Make sure they all follow the same process for measurement – ideally written down. For example, make sure they all measure the parts in the same place. This will reduce the "within part" variation.
- Always select enough parts to represent the entire process variation. If you don't do this then you are likely to over-estimate the proportion of measurement error in the system.
- Take the measurement in random order so that the people doing the measuring have no knowledge of previous measurements. This prevents them "influencing" the results such as varying the tightness of a micrometer to enable the same repeated result.

There are three R&R designs available to us and these are:

A Crossed Design – used when multiple parts are measured by multiple inspectors a number of times.

A Nested Design – used when each part is measured by only one inspector such as when the parts are subject to destructive test and therefore not available to be measured by another inspector.

An Expanded Design – commonly used when you are assessing more than simply the inspectors and parts.

The vast majority of Gage R&R studies simply want to assess the measurement variation due to parts and inspectors and it is quite rare to need a nested or expanded design. We will therefore concentrate on the crossed design.

Let's assume that we have considered the notes described above and decided to use 10 sample parts, 3 inspectors and two replicates. We believe that the 10 parts represent the total variation of the process and that the inspectors are highly skilled and therefore the variation between them is small.

2. Go to Stat/Quality Tools/Gage Study/Create Gage R&R Study Worksheet

3. You'll see this entry screen:

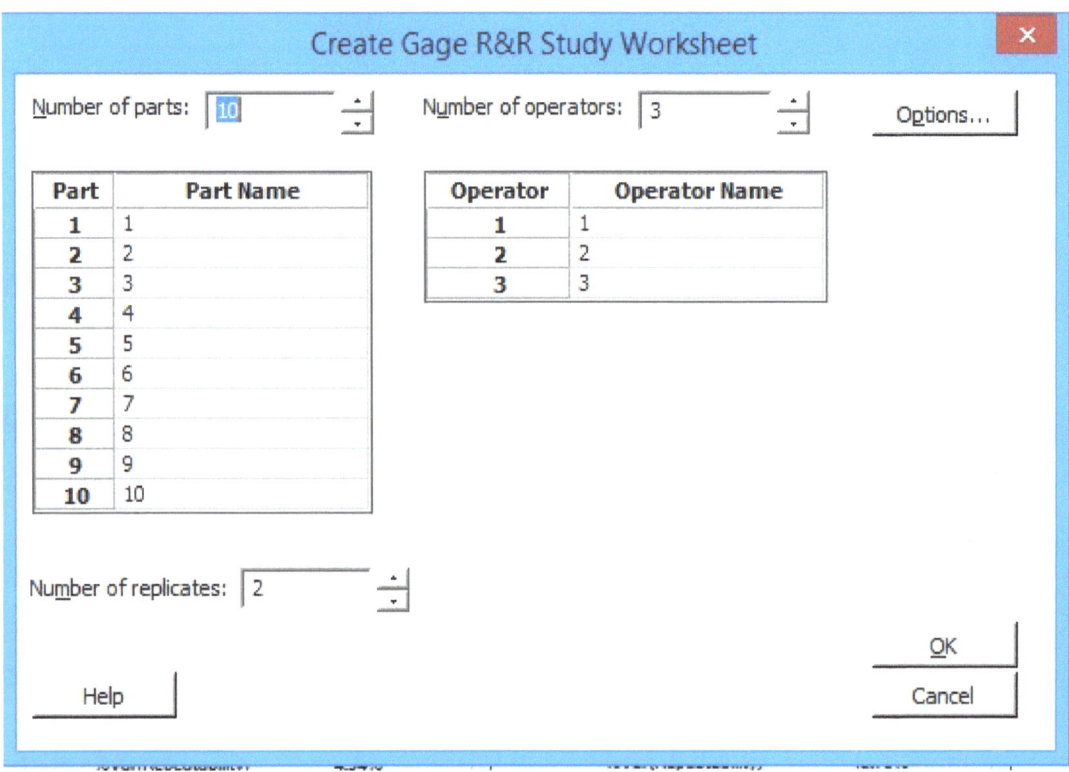

Enter the number of parts, number of inspectors (called operators by Minitab) and the number of replicates. You'll see that in the top right of the screen is an options button. If you click this you'll see this screen:

Create Gage R&R Study Worksheet: Options

- ○ Do not randomize
- ○ Randomize all runs
- ● Randomize runs within operators
  - ☐ Randomize operator sequence

Base for random data generator: [ ]

☐ Store standard run order in worksheet

[Help] [OK] [Cancel]

Let's examine these in turn so that we understand them.

The first three radio buttons give you options for randomization. Wherever possible you should use some level of randomization. Choose as follows:

"Randomize all runs" when you want to randomize the run order of all the measurements. Use this where possible.

"Randomize runs within operators" when there is a reason for each inspector to make measurements of all parts and replicates before any other inspector does any measurements. It might be, for example, that the inspection sequence will take several hours and therefore you want to release the inspectors back to production when they have completed the measurements. In this case randomizing runs within operators will enable you to release each operator in turn as soon as they have completed their measurements.

"Randomize operator sequence" only appears if you choose "randomize runs within operators". Use this if you want the 2nd inspector to measure runs 1-10, the first operator to measure runs 11-20 etc.

It is possible to set a base for the random data generator and this will generate identical random orderings every time the same base is used.

Finally there is an option to store the standard run order in the worksheet.

In this case there is no reason not to randomize all runs so we check this radio button and leave all other options blank. We then press "Ok" to return to the Create Gage R&R Study Worksheet. Click on "Ok" on this sheet and Minitab will tabulate the experiment to be carried out, listing the Run Order, the Parts and the Operators.

Now carry out the experiment, completing each run in turn. The results (called "Responses" by Minitab) are placed in the column next to "Operator".

Next it's time to carry out the analysis and interpret the results.

Go to Stat/Quality Tools/Gage Study/Gage R&R Study (Crossed)

You will be presented with the following input screen:

The select button may be greyed out. To enable it simply click on C1, C2 or C3. You will now be able to select each of these in turn into "Part numbers", "Operators" and "Measurement Data" respectively.

You have two choices for "Method of Analysis"; ANOVA and Xbar and R. So which one should you choose?

Well, both these are just methods of breaking down the overall variation into its component parts. The Xbar and R option looks at part to part variation and gage repeatability and reproducibility. The ANOVA method does the same but also breaks down the reproducibility into operator and operator by part.....so it looks into slightly more detail than the Xbar and R method.

I would generally therefore use the ANOVA method since it's actually no extra work to do so.

Let's look at an example using the Minitab sample data. The file name we are going to use is

GAGEAIAG.MTW

You can either use the copy included in Minitab (it's in the sample data file) or download a copy from:

http://support.minitab.com/en-us/datasets/

Open this dataset in Minitab.

In this example, ten parts are measured three times by three different inspectors. This is done in a random order. The ten parts are selected to cover the entire range of variability.

To carry out the analysis:

Choose Stat/Quality Tools/Gage Study/Gage R&R Study (Crossed)

Select the various options as follows:

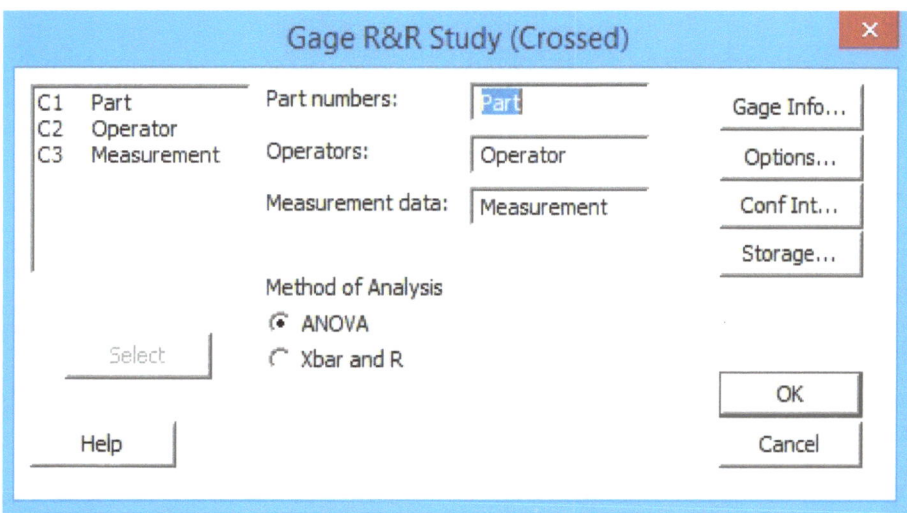

Click on the "Options" button near top left and under "Process tolerance" select Upper spec – Lower spec and enter the number 8 so it looks like this:

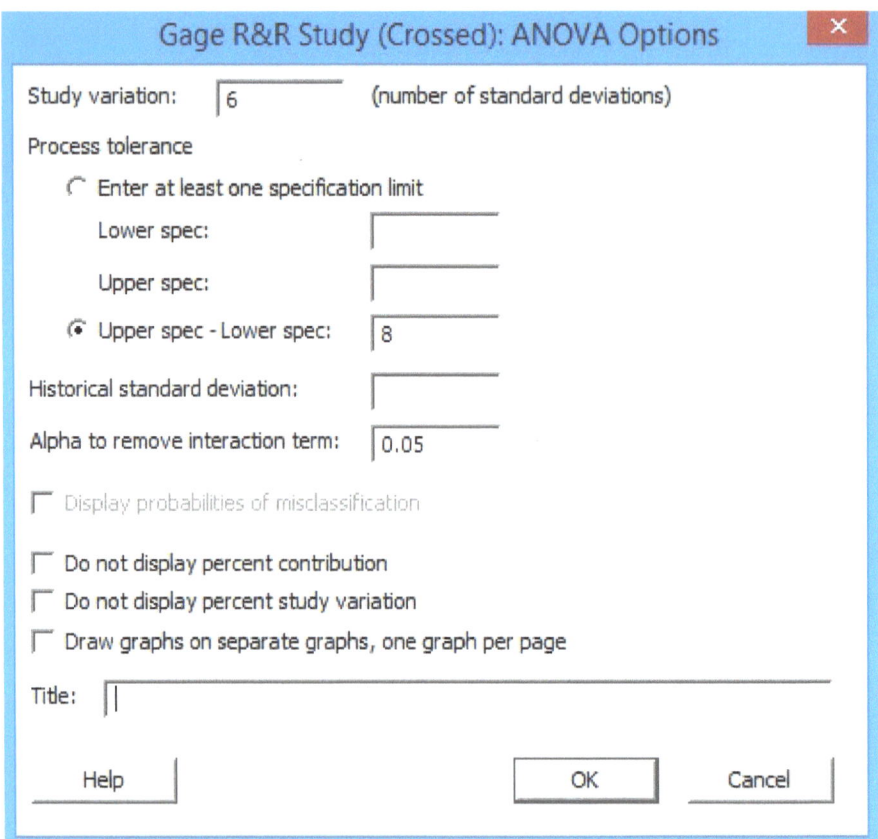

Do not alter any other settings.

Now click Ok in each dialogue box.

You'll see several tables displayed in the session window and six graphs displayed in the graph window. Let's look at how to interpret the results.

The whole idea of this analysis is to ensure that the part to part variation is significantly greater than the variation due to the measurement system.

I find that this simple question can quickly be answered by looking at the six graphs:

Gage R&R (ANOVA) Report for Measurement

Gage name:
Date of study:

Reported by:
Tolerance:
Misc:

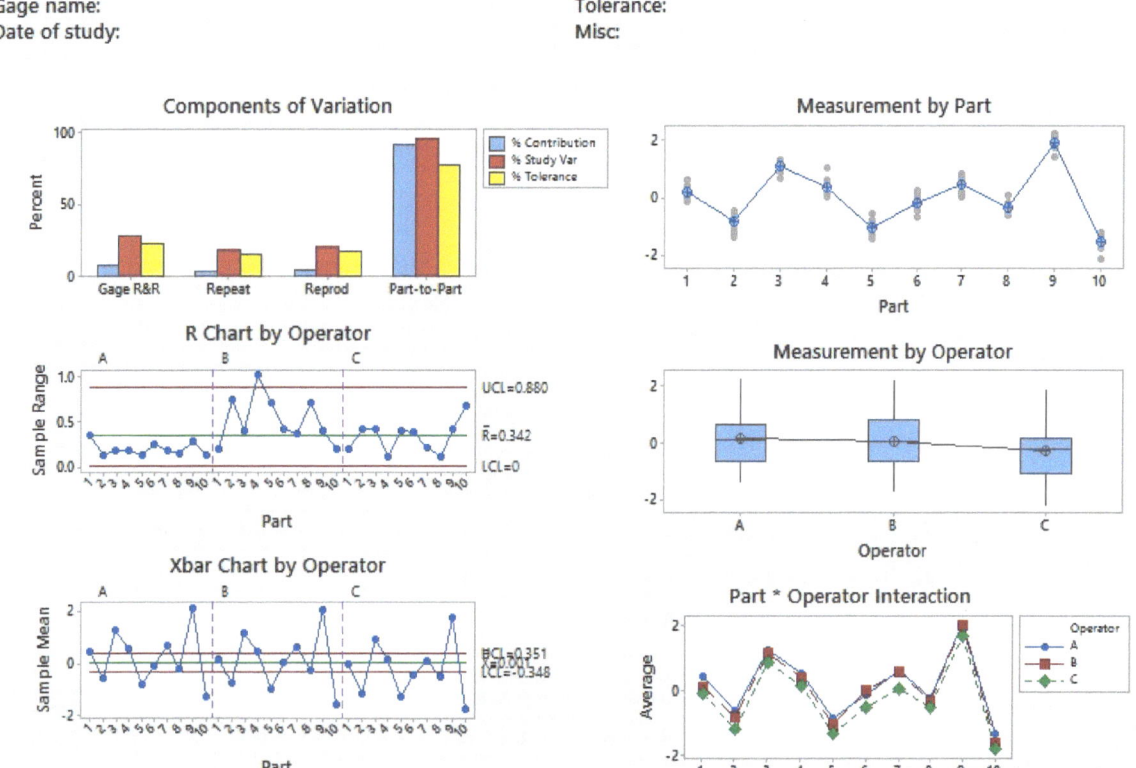

In the graph in the top left corner we can see that the part to part variation is much greater than the total gage R&R – so to begin to answer the question, it certainly appears that the majority of the variation is due to the differences between the parts.

In the graph in the upper right corner, this deduction is substantiated by the fact that the line varies and is not flat.

In the graph in the middle left position, we can see that inspector B (the data in the middle block of the graph) has measurements that are much more variable than her two colleagues.

In the graph in the middle right position the difference between inspectors is again small when compared with the difference between parts.

In the graph in the bottom left position, the points on the chart generally fall outside the control limits which again tells us that the variation is mostly due to the differences between the parts.

The graph on the bottom right shows the interaction between parts and inspectors.....and in this case the tables in the session window tell us that the value is

0.974, meaning there is no significant interaction between the inspector and each part. You can see this highlighted yellow in the table below.

Gage R&R Study - ANOVA Method

Two-Way ANOVA Table With Interaction

Source	DF	SS	MS	F	P
Part	9	88.3619	9.81799	492.291	0.000
Operator	2	3.1673	1.58363	79.406	0.000
Part * Operator	18	0.3590	0.01994	0.434	0.974
Repeatability	60	2.7589	0.04598		
Total	89	94.6471			

Let's take a look to see if the measuring system is acceptable. There are two sets of rules:

The total gage R&R percentage in the % Study Variation column should ideally be less than 10%. It may be acceptable at up to 30% depending on the cost of the measuring device and the application.

The total gage R&R percentage in the % Contribution column should ideally be less than 1% but may be acceptable at up to 10% depending on the cost of the measuring device and the application.

In this case

- The total gage R&R in the contribution column is 7.76% and the part to part variation is 92.24% so we can conclude that the measurement system may be acceptable but it certainly could be improved.

Gage R&R

Source	VarComp	%Contribution (of VarComp)
Total Gage R&R	0.09143	7.76
Repeatability	0.03997	3.39
Reproducibility	0.05146	4.37
Operator	0.05146	4.37
Part-To-Part	1.08645	92.24
Total Variation	1.17788	100.00

- The total gage R&R in the % study variation column is 27.86% and the part to part variation is 78.17% % so we can conclude that the measurement system may be acceptable but it certainly could be improved.

Source	StdDev (SD)	Study Var (6 × SD)	%Study Var (%SV)	%Tolerance (SV/Toler)
Total Gage R&R	0.30237	1.81423	27.86	22.68
Repeatability	0.19993	1.19960	18.42	14.99
Reproducibility	0.22684	1.36103	20.90	17.01
Operator	0.22684	1.36103	20.90	17.01
Part-To-Part	1.04233	6.25396	96.04	78.17
Total Variation	1.08530	6.51180	100.00	81.40

Ok, that's all I want to cover in step 2. Basically all you are looking to do is ensure that the measurement system itself is fit for purpose.

This means that the measurement system should contribute relatively little variation when compared with the parts or parameters being measured.

It's now time to get into the production of the actual control charts.

Step 3: Choose Appropriate Control Chart Type and Decide on Reaction Plan

If you have the latest version of Minitab, which at the time of writing is version 17, then you have access to the Assistant.

To see this, click on the "Assistant" tab at the top right of the top of the page and then select "control charts" at the bottom of the list.

You'll then see a selection tree to help you make a decision. It looks like this:

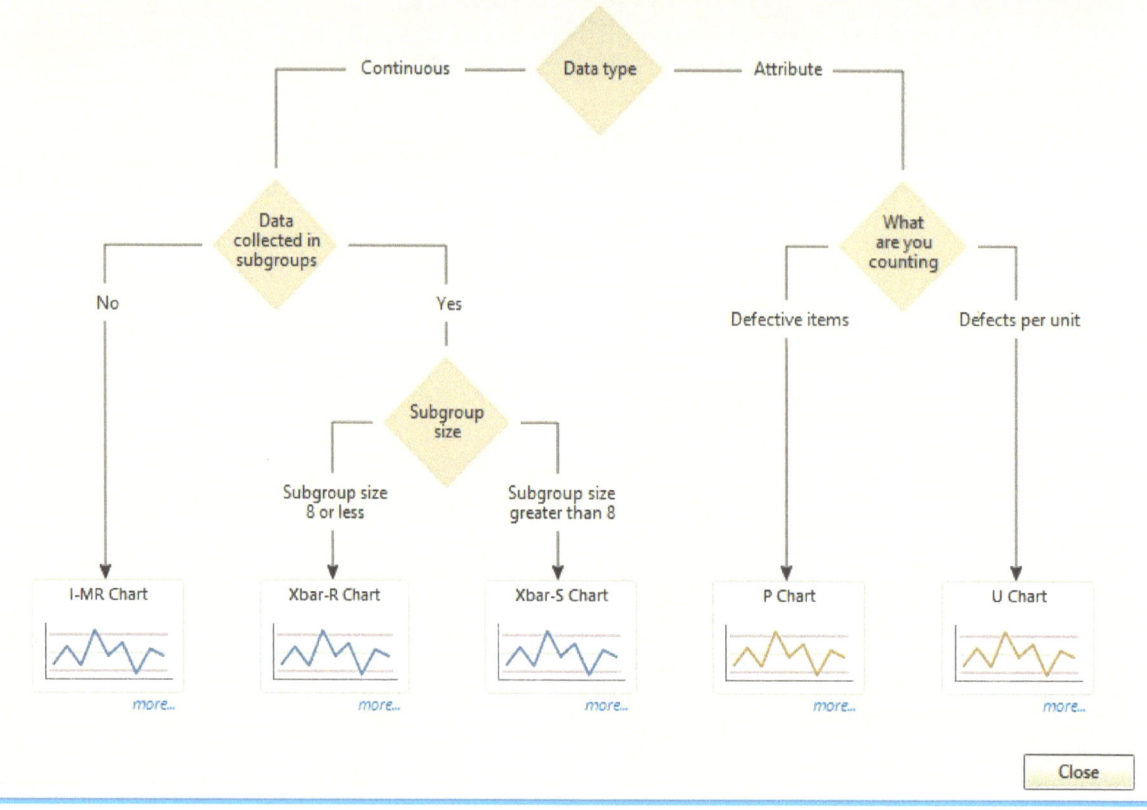

You'll see that the first decision is based on the data type being collected.

There are 5 types of control chart, 3 of them for continuous data and two of them for attribute data. The types are:

For continuous data;

- I-MR Chart
- Xbar-R Chart
- Xbar-S Chart

For attribute data;

- P Chart
- U Chart

My strong advice is to always use continuous data rather than attribute data. This type of data will yield far more information and require less results to be meaningful.

If you click on the diamond which says "Data collected in sub-groups" you'll see some guidance that explains what is meant by sub-groups. It looks like this:

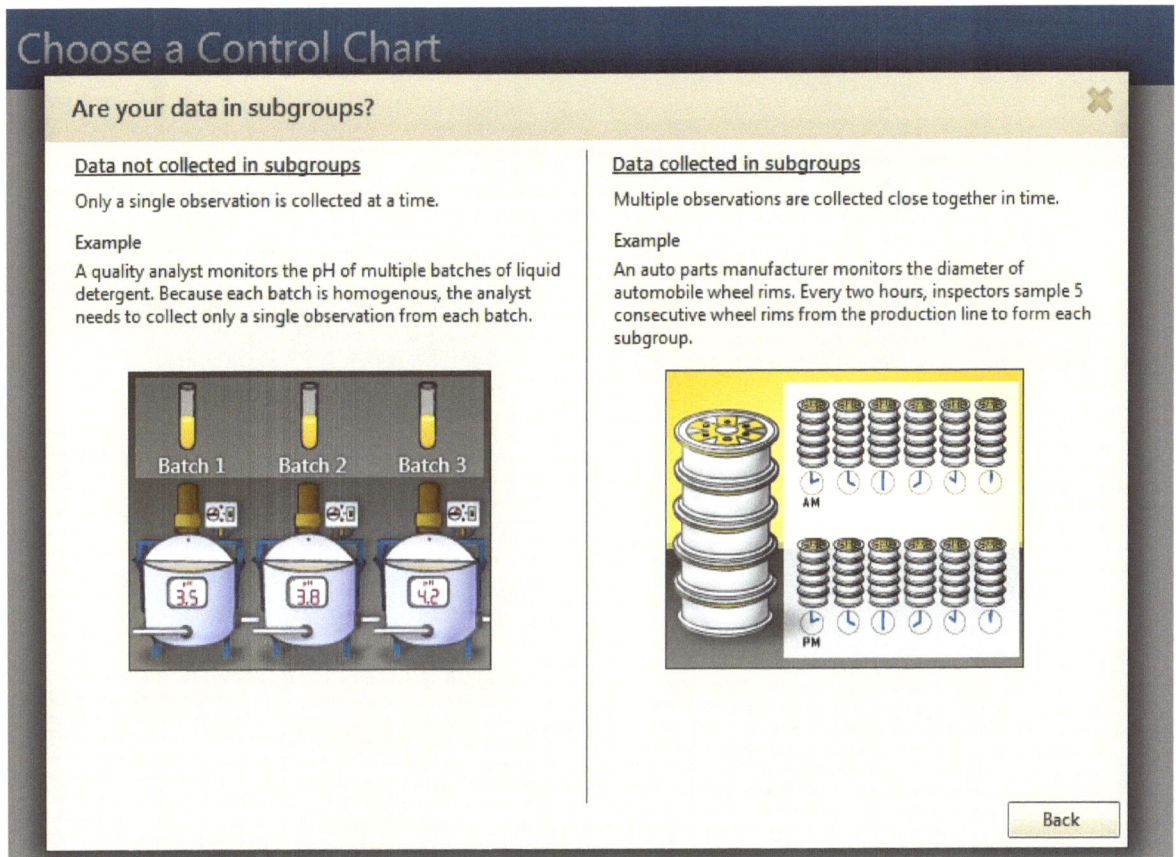

The difference is shown with a couple of examples and will be dependent on the types of samples being taken. Let's assume for this example that we are going to collect data in sub-groups and therefore we follow the leg marked "Yes" and reach the diamond marked "Subgroup size". Again click on this diamond and you'll see more guidance like this:

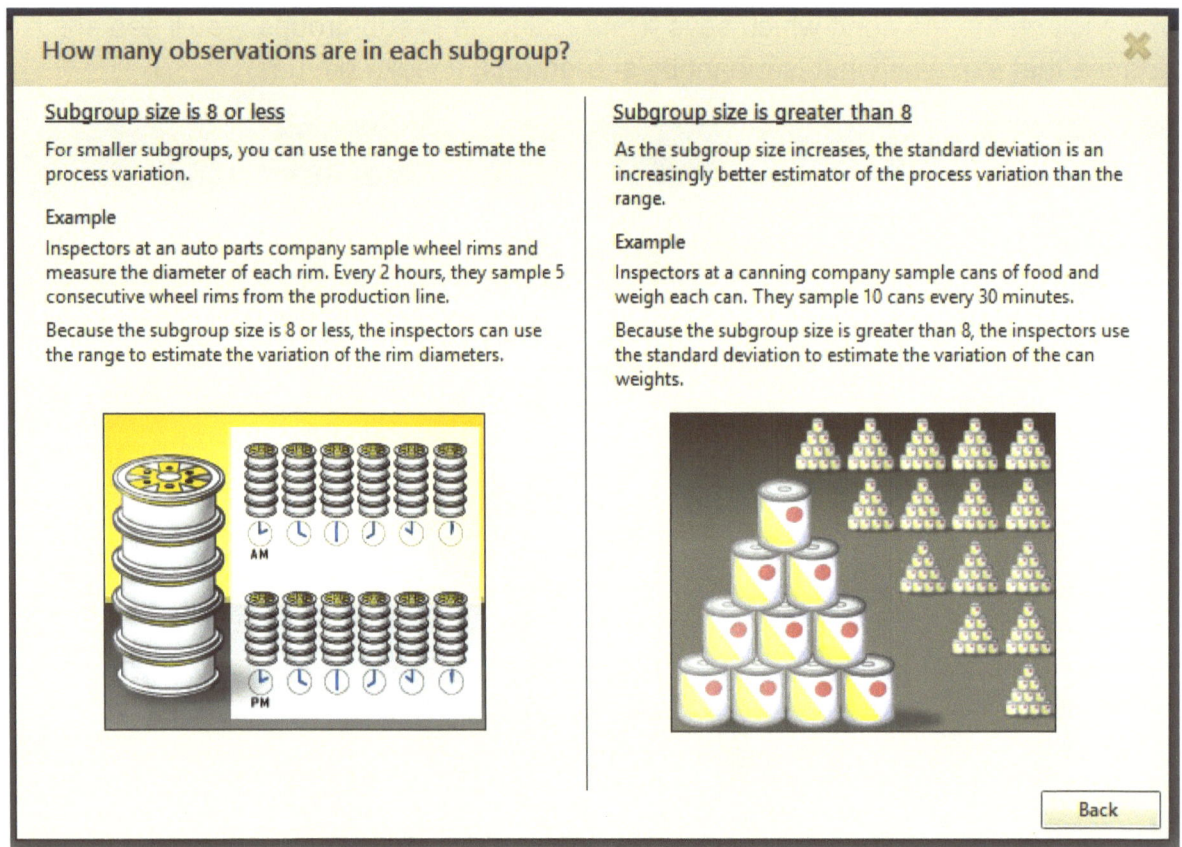

Clearly the larger the sub group size, the better the estimate of process variation but this will to be balanced against the cost and time to take the samples and measure them. We'll choose a sub-group size of 10 which is obviously greater than 8 so we therefore follow this route on the decision tree and we end up at the x-bar-S chart.

To select the type of chart you are going to use it's simply a matter of following the Assistant to arrive at the correct chart.

Step 4: Collect the Data

The key point when taking the sub groups is to make sure that each sub group is collected under the same conditions. This means the same machine, same producer, same shift......as much as possible the same conditions for each sub group.

The sub groups should be taken at regular intervals and you have to decide what interval is required. It should be short enough to capture changes in variation of the process. In many cases every 30 minutes or every hour is sufficient but use judgement.

Since we are going to use the data collected to calculate the control limits it is essential that we use enough data to form an estimate that is precise......in reality you'll need at least 100 data points. Minitab will check to see if enough data points exist to give a precise estimate.

Let's run through an example to demonstrate. This example is provided in the Minitab Tutorials but I'll walk you through it and explain the key points.

This is the example as described directly by Minitab:

"A liquid detergent company wants to assess whether its process is in control. The company makes the liquid detergent in batches by mixing together a number of ingredients. The quality characteristic of interest is the pH value for each batch. Measurements of the pH value for 25 consecutive batches were taken"

This is a typical scenario in many companies. I did a very similar analysis for an explosives mixing plant a few years ago!

In this example only one sample was taken for each batch and therefore each sample is considered to be an independent observation – this means no sub-groups.

We are going to use the Minitab supplied sample data called "DETERGENT.MTW" which you can find in the Minitab sample data folder. You can reach this by clicking on File/Open Worksheet. You'll then see this:

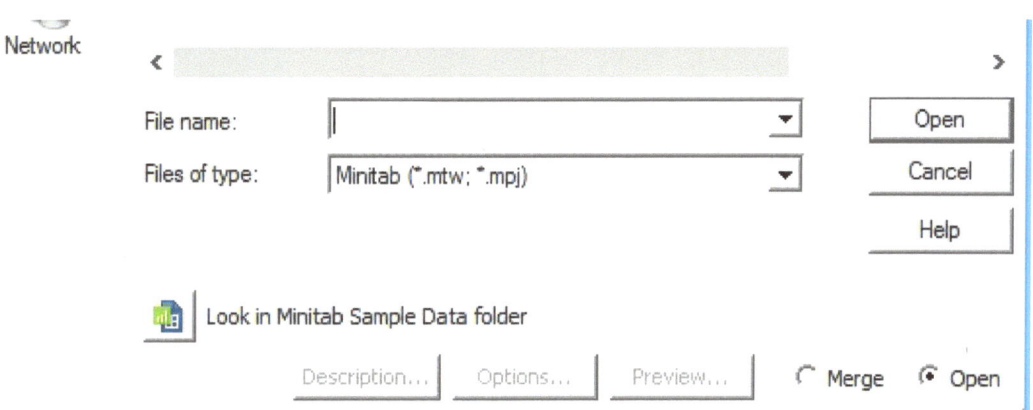

Now click on the button marked "Look in Minitab Sample Data folder" and select DETERGENT.MTW

You'll see a message saying that a copy of the file will be added to the current project and column C1 will be populated with the detergent data.

Now choose Stat/Control Charts/Variable Charts for Individuals/I-MR

You'll then be presented with an input screen like this:

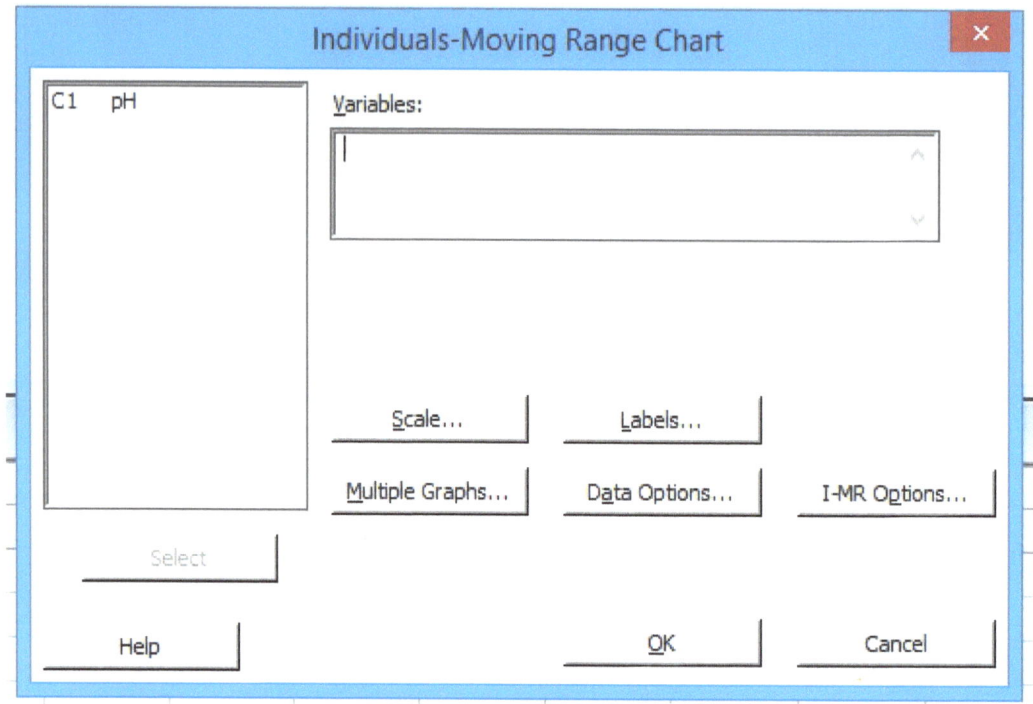

Click on the C1 variable in the left hand side box and click on the button marked "Select". This will place the variable into the "Variables" box.

There are 5 option buttons marked:

"Scale"

"Labels"

"Multiple Graphs"

"Data Options"

"I-MR Options"

Here's what they mean:-

"Scale" is used to specify a time scale for the x-axis. You can edit this after creating the graph. You can edit the range, the tick labels, attributes, font, alignment, and display of the scale. You can also use something called the "stamp" option. This allows you to label the tick marks on the x-axis.

"Labels" allows you to edit the title and sub-titles of the graph.

"Multiple Graphs" is simply used to control the placement and scales of multiple control charts.

"Data Options" allows you to include or exclude certain values or rows in the data.

"I-MR Options" contains 8 input screens:

"Parameters" allows you to enter the mean and standard deviation of the parameter rather than letting Minitab calculate them from the data.

"Estimate" allow you to omit or include certain parameters based on pre-assigned rules – no need to modify this.

"Limits" allows you to draw control limits above and below the mean at the multiples of any standard deviation. Again no need to enter anything here.

"Tests" allows you to select any or all of the eight tests for special causes. These eight tests look for specific patterns in the plotted control chart and if it finds it considers it to be special cause meaning it's not random variation in the process but something else acting on the system.

"Stages" allows you to draw a historical control chart to see the difference in data both before and after a change has been made.

"Box-Cox" is a way of transforming data which is not normal to make it more normal. The data we are using here for this example is normal so no need to use a Box-Cox transformation.

"Display" allows you to alter the way the control chart is displayed.

"Storage" allows you to store specific statistics in the actual worksheet.

Ok, so we not going to modify any of these options, and we go ahead and press "Ok" on the input screen. The following graph will be produced.

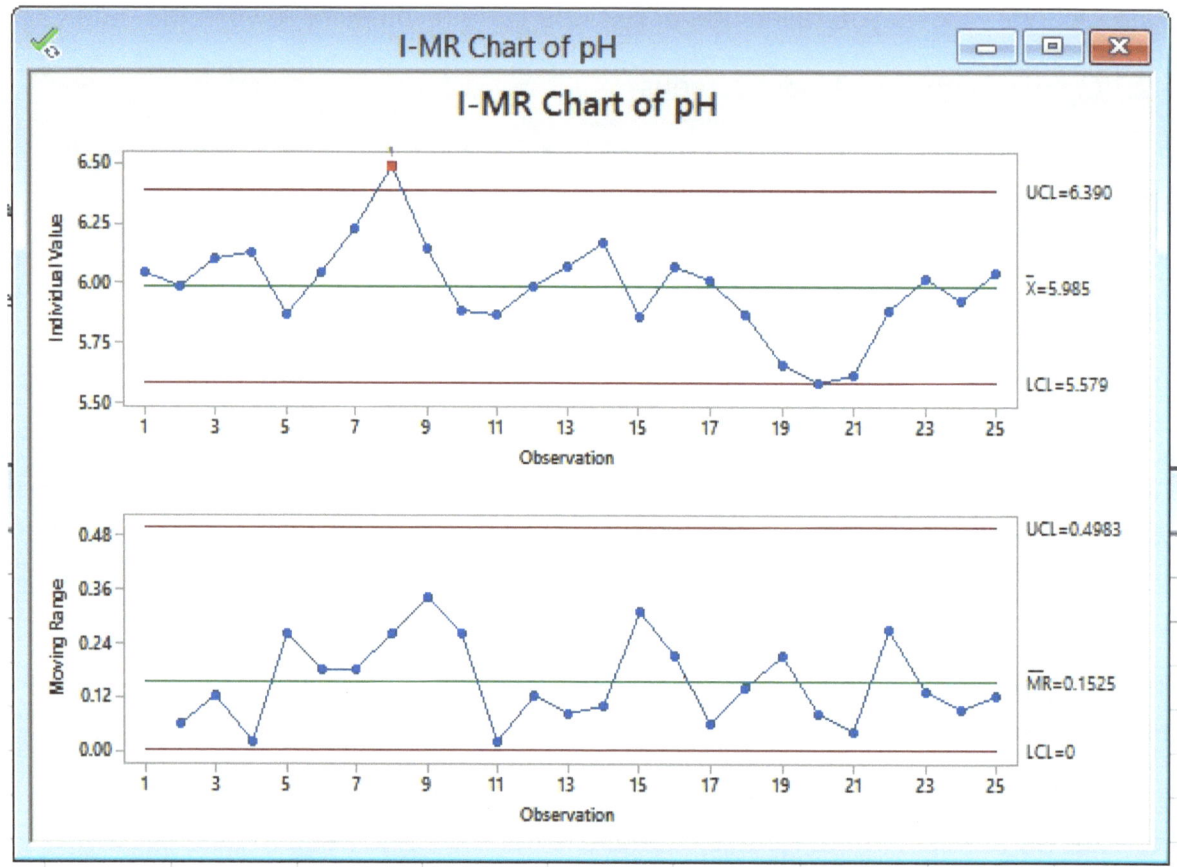

It looks nice but how do we interpret it and what does it mean?

Step 5: Calculate the Control Limits and Decide if Process Is In Control

Let's look at the bottom chart, which is called a "moving range chart". Minitab checks the data and tests for special causes, which simply means that it checks to see if the plotted points are distributed randomly within the control limits.

You can see in this case that the Upper Control Limit (UCL) on the chart has been calculated at 0.4983 and that all data points are within this limit. This means the data is in control and therefore you can go on to look at the individual value chart. If Minitab had detected any out of control conditions (i.e. outside the control limits) then it's possible that a special cause exists and it should be investigated.

Turning our attention to the individuals chart we can see that the output from Minitab tells us that there is one point which is more than 3 standard deviations from the mean (point 8) and this is clearly marked on the chart.

Minitab also reports that points 20 and 21 are 2 out of 3 continuous points which are more than 2 standard deviations from the green centre line (an estimate of the process average). This means there MAY be special causes present which could be investigated.

This is where people simply assume that because Minitab has found some data points which fail the various tests for special causes, then it must be present and should always be investigated. This is not so. Remember that we left the settings such that Minitab is using all eight tests for special causes and my experience is that many data sets appear to indicate special causes when all eight tests are used.

So what do we do in this case? Well carry out some preliminary investigation to check the data point which appears outside the control limit. Could there be a logical reason for it. Reasons include:

A different piece of measuring kit being used (maybe the one used for all the other data points was unavailable for some reason?

A different person took this one measurement?

You are looking for an explanation but you may not find one and therefore have to decide on the data you have if special causes are present. In this case the data point is 6.49 against a control limit of 6.39.

Will this affect your conclusions? I doubt it. In this example I would conclude from looking at both graphs that the process is in control and I can therefore rely on the data.

Step 6: Determine if Data is Normally Distributed

We now need to determine if the data is normally distributed. Some analyses (such as a capability analysis) could mislead you if the data is non normal so it's worth knowing if the data is normal.

Minitab is able to carry out a test for normality. Go to Stat/Basic Statistics/Normality Test and you'll see this screen:

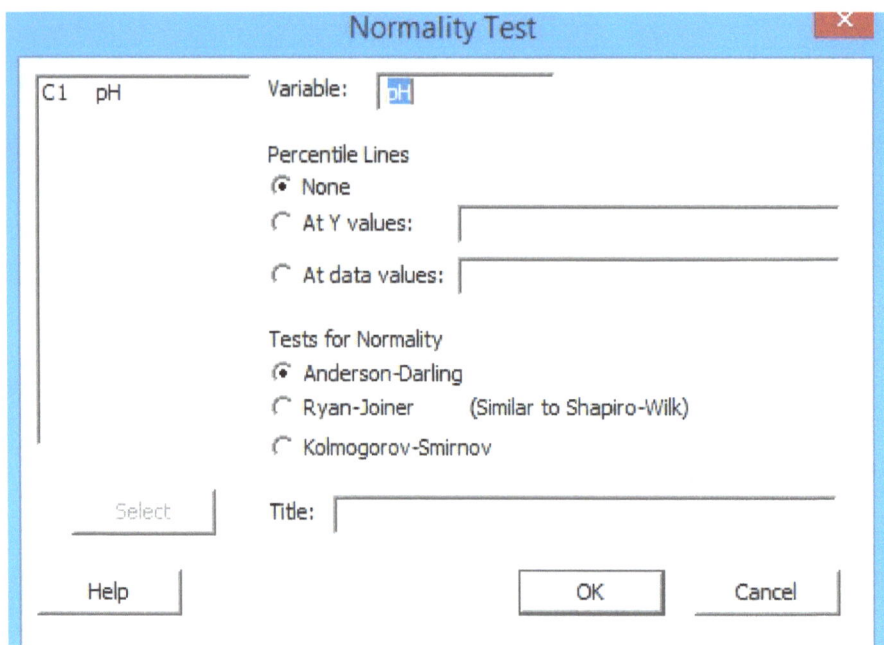

Click on the variable C1 in the left hand box and it will place it in the box marked "Variable".

There is now several options:

"Percentile Lines", with a choice of "None", "At Y values" or "At data values"

"Tests for Normality" with a choice of "Anderson-Darling", "Ryan-Joiner" and the brilliantly named "Kolmogrov - Smirnov.

Let's see what these mean.

"Percentile lines" means that Minitab marks each percent in the data with a corresponding horizontal reference line on the plot. I sometimes use this when analysing lots of data just to make it clearer to see the various data point. If you choose "At Y values" you can enter the y-scale values for positioning the percentile lines or you can use the actual data values for positioning them. In reality I seldom use either since I don't think they add much clarity (if any) to the analysis.

"Tests for Normality" – although you have three choices, which one should you use?

Well the short answer is that it seldom matters too much and I have a cheap and easy way of checking for normality using a piece of equipment that almost everyone in the world has. Let's produce the plot and I'll demonstrate.

Leave the option for percentile lines at "None" and the normality test as "Anderson-Darling". Now click on "Ok". Minitab will then carry out the test for normality and produce the plot which will look like this:

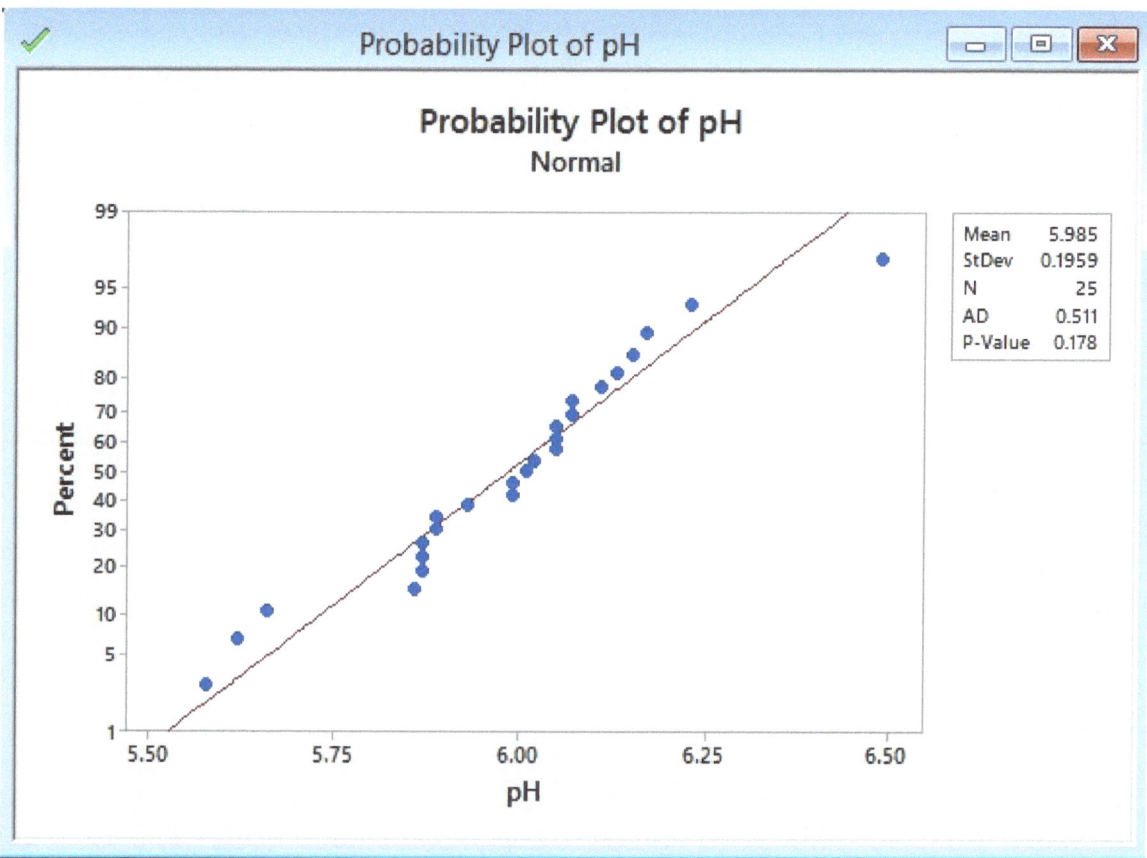

Now I use my simple test.

Can I place a straight finger across the data and cover all the data points? If the answer is "yes" then it's near enough normal to make little difference in the conclusions you draw from the data set. It's only a very rough test but good enough in most cases to assess the normality of data. This infuriates the purists who insist on carrying the above tests but I've generally found it to be "good enough".

In this case I am satisfied that this is data which is normally distributed.

Step 7: Decide What Action Is Required to Reduce Variation

This is probably the most important step.

The whole idea of statistical process control is to assess the capability of a process and provide evidence of when a change or alteration should be made.

This is one of the most common mistakes when people start to collect data and try and assess it. They often conclude that there is a pattern in the data, when actually none exists and also try to control what is normal variation in the process. THIS OFTEN DOES MORE HARM THAN GOOD!

Be sure that you have the required evidence before you start adjusting a process. In many cases such adjust becomes continual as you chase the variation and the variation actually increases, I have seen this happen many times over the years!

So, what action is required, if any?

Well, if you have clear evidence of special cause variation then you should investigate since you will find it impossible to predict the performance of the process and it will often surprise you and most likely cost time and money.

The variation might show no special causes but the tolerance sit inside the control limits. If this is the case then a proportion of what is produced will be "out of spec" and Minitab will be able to show you an estimate of just how many.

If the tolerances lie outside the control limits then you have good control of the process and if the limits have been calculated as 3 sigma limits then 99.73% of items produced will be "in spec". This may or may not be good enough. What does the customer expect?

In many cases the customer will simply say "I want them all to be within specification" but in reality this can never be an agreed acceptance standard. Does it mean one in 100 can be outside specification? Or 1 in 1000? Or 1 in 10,000? Or 1 in 1000000?

The level of work to provide the control and evidence for each of these standards is wildly different. Always understand and have an agreed acceptance standard with your customer.

It's exactly the same with suppliers. Give them an acceptance standard – what proportion of items are you prepared to accept that are out of specification? How will they demonstrate that they have reached this level?

The best suppliers will take a sample of sufficient size and use control charts to demonstrate control and compliance. Note that a control chart provides both these aspects.

Control because they will know when a process changes (or is about to change) and therefore they can do something about it. Compliance because the control chart itself is the evidence that they meet the specification.

If changes are required to the process then other techniques can be used to see where the variation is originating and what factors have the most impact. One of the most powerful is called "Design of Experiments", or DoE for short. To see how this should really be done grab a copy of my book:

"Practical Design of Experiments" available through Amazon.

Ok, that's practical Statistical Process Control.

Done correctly it will save you a fortune. Incorrectly done and it will cost you time, money and credibility.

Finally if you wish to take advantage of my skills and want me to improve your business and its bottom line then you can reach me via my consultancy company at:

http://hardwickconsulting.co.uk

Good luck!

Glossary of Terms

Measurement System Analysis: A method of qualifying a measurement system before use to ensure it is precise, accurate and stable.

Statistical Process Control: A method of assessing process data which relies on statistics to inform about process performance.